Firefighter Fatalities
in the
United States
in 1996

Prepared for

United States Fire Administration
Federal Emergency Management Agency
Contract No. EMW-95-C-4713

Prepared by

TriData Corporation
1000 Wilson Boulevard
Arlington, Virginia 22209

August 1997

Table of Contents

ACKNOWLEDGMENTS

This study of firefighter fatalities would not have been possible without the cooperation and assistance of many members of the fire service across the United States. Members of individual fire departments, chief fire officers, the National Interagency Fire Center, US Forest Service personnel, the US military, the Department of Justice, and many others contributed important information for this report.

TriData Corporation of Arlington, Virginia conducted this analysis, for the United States Fire Administration under contract EMW-95-C-4713.

The ultimate objective of this effort is to reduce the number of firefighter deaths through an increasing awareness and understanding of their causes and how they can be prevented. Firefighting, rescue and other types of emergency operations are essential activities in an inherently dangerous profession, and tragedies will occur from time to time. This is the risk all firefighters accept every time they respond to an emergency incident. However, the risk can be greatly reduced through efforts to increase firefighter health and safety.

The United States Fire Administration would like to extend its thanks to the Boston Fire Department's Public Relations Office for providing the photograph for the cover. The picture was taken during the dedication ceremony for the Vendome Memorial in Boston, MA. The Hotel Vendome fire occurred on June 17th, 1972 and killed nine Boston firefighters.

This report is dedicated to those firefighters who have made the ultimate sacrifice in 1996. May the lessons learned from their passing not go unheeded.

BACKGROUND

For the last 20 years, the United States Fire Administration (USFA) has tracked the number of firefighter fatalities and conducted an annual analysis. Through the collection of information on the causes of firefighter deaths, the USFA is able to focus on specific problems and direct efforts towards finding solutions to reduce the number of firefighter fatalities in the future. This information is also used to measure the effectiveness of current programs directed toward firefighter health and safety.

In addition to the analysis, the USFA maintains a list of firefighter fatalities for the Fallen Firefighter Memorial Service. The fallen firefighters' next of kin as well as members of the individual fire departments are invited to the annual Fallen Firefighter Memorial Service, which is held at the National Fire Academy in Emmitsburg, Maryland every fall. Additional information regarding the memorial service can be found on the Internet at http\\www.usfa.fema.gov or by calling the National Fallen Firefighters Foundation at 301-447-1365.

INTRODUCTION

This report continues a series of annual studies by the US Fire Administration of firefighter fatalities in the United States.

The specific objective of this study was to identify all of the on-duty firefighter fatalities that occurred in the United States in 1996, and to analyze the circumstances surrounding each occurrence. The study is intended to help identify approaches that could reduce the number of firefighter deaths in future years.

In addition to the 1996 overall findings, this study includes special analyses on violent firefighter deaths, physical fitness and its relation to firefighter deaths, and vehicle accidents.

Who Is a Firefighter?

For the purpose of this study, the term *firefighter* covers all members of organized fire departments, including career and volunteer firefighters; full-time public safety officers acting as firefighters; state and federal government fire service personnel; including wildland firefighters; and privately employed firefighters, including employees of contract fire departments and trained members of industrial fire brigades, whether full or part-time. It also includes contract personnel working as firefighters or assigned to work in direct support of fire service organizations.

Under this definition, the study includes not only local and municipal firefighters, but also seasonal and full-time employees of the United States Forest Service, the Bureau of Land Management, the Bureau of Indian Affairs, the Bureau of Fish and Wildlife, the National Park Service, and state wildland agencies. It also includes prison inmates serving on firefighting crews; firefighters employed by other governmental agencies such as the United States Department of Energy; military personnel performing assigned fire suppression activities; and civilian firefighters working at military installations.

What Constitutes an On-Duty Fatality?

On-duty fatalities include any injury or illness sustained while on-duty that proves fatal. The term *on-duty* refers to being involved in operations at the scene of an emergency, whether it is a fire or non-fire incident; being en route to or returning from an incident; performing other officially assigned duties such as training, maintenance, public education, inspection, investigations, court testimony and fund-raising; and being on-call, under orders, or on stand-by duty, except at the individual's home or place of business.

These fatalities may occur on the fireground, in training, while responding to or returning from alarms, or while performing other duties that support fire service operations.

A fatality may be caused directly by accident or injury, or it may be attributed to an occupational-related fatal illness. A common example of a fatal illness incurred on duty is a heart attack. Fatalities attributed to occupational illnesses would also include a communicable disease contracted while on duty that proved fatal, where the disease could be attributed to a documented occupational exposure.

Accidents that claim the lives of on-duty firefighters are also included in the analysis, whether or not they are directly related to emergency incidents. In 1996, this category includes a firefighter who died in a car accident while in transit between fire station assignments and a total of six firefighters who were victims of violence against emergency service personnel.

Injuries and illnesses are included where death is considerably delayed after the original incident. When the incident and the death occur in different years, the analysis counts the fatality as having occurred in the year that the incident occurred. For example, a firefighter died in 1996 of medical complications that resulted from an injury at a fire in 1982. Because his death was the result of the 1982 incident, this case was counted as a 1982 fatality for statistical purposes, and is not included in the 94 fatalities for 1996 that were analyzed in this report. Since the death occurred in 1996, he will be included in the 1996 annual Fallen Firefighter Memorial Service at the National Fire Academy, and his name will be included on the list of firefighters who died in 1996.

There is no established mechanism for identifying fatalities that result from illnesses that develop over long periods of time, such as cancer, which may be related to occupational exposure to hazardous materials or products of combustion. It has proven to be very difficult over several years to provide a full evaluation of an occupational illness as a causal factor in firefighter deaths, because of the limitations in the ability to track the exposure of firefighters to toxic hazards, the often delayed long-term effects of such exposures, and the exposures firefighters may receive while off-duty.

Sources of Initial Notification

As an integral part of its ongoing program to collect and analyze fire data, the United States Fire Administration solicits information on firefighter fatalities directly from the fire service and from a wide range of other sources. These include the Public Safety Officer's Benefit Program (PSOB) administered by the Department of Justice, the Occupational Safety and Health Administration (OSHA), the US military, the National Interagency Fire Center, and other federal agencies.

The USFA receives notification of some deaths directly from fire departments, as well as from fire service organizations such as the International Association of Fire Chiefs (IAFC), the International Association of Fire Fighters (IAFF), the National Fire Protection Association (NFPA), the National Volunteer Fire Council (NVFC), state fire marshals, state training organizations, other state and local organizations, and fire service publications. The USFA also keeps track of fatal fire incidents as part of its Major Fire Investigations Project and maintains an ongoing analysis of data from the National Fire Incident Reporting System (NFIRS) for the production of the report *Fire in the United States.*

Procedure for Including a Fatality in the Study

In most cases, after notification of a fatal incident, initial tele   phone contact is made with local authorities by the USFA's contractor to verify the incident, its location and jurisdiction, and the fire department or agency involved. Further information about the deceased firefighter and the incident may be obtained from the chief of the fire department or his designee over the phone or by other data collection forms.

Information that is routinely requested includes NFIRS-1 (incident) and NFIRS-3 (fire service casualty) reports, the fire department's own incident reports and internal investigation reports, copies of death certificates or autopsy results, special investigative reports such as those produced by the USFA or NFPA, police reports, photographs and diagrams, and newspaper or media accounts of the incident.

After obtaining this information, a determination is made as to whether the death qualifies as an on-duty firefighter fatality according to the previously described criteria. The same criteria was used for this study as in previous annual studies. Additional information may be requested, either by follow-up with the fire department directly, or from state vital records offices or other agencies. The determination as to whether a fatality qualifies as an on-duty death for inclusion in the statistical analysis and the Fallen Firefighter Memorial Service is made by the USFA.

1996 FINDINGS

Ninety-four (94) firefighters died while on duty in 1996.[1] This is a slight decrease from last year's total of 96. The total of 94 fatalities is the third lowest number recorded in the 20 years that this data has been collected, and is only the fourth time that the total has been less than 100 fatalities. The lowest years were 1992 with 75 fatalities and 1993 with 77 fatalities.

This year's total is part of a long-term downward trend of reduced fatalities that began in 1979 after a peak of 171 in 1978. The overall trend in firefighter fatalities is down 35 percent over the last ten years. Over the last five years there has been an upward trend of 29 percent, though the number of deaths in 1996 decreased approximately two percent from 1995 (Figure 1).

Figure 1. On-Duty Firefighter Deaths for 1996

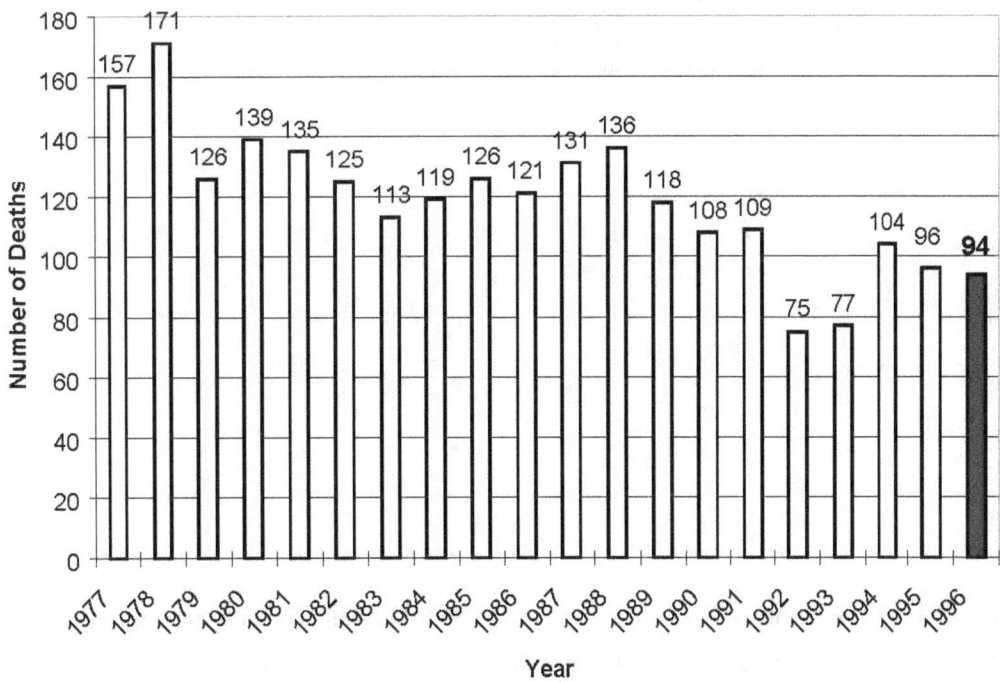

---

[1] As mentioned earlier, the 94 on-duty fatalities in 1996 do not include one firefighter who died during the year from injuries sustained in 1982. This firefighter was injured when a concrete loading dock collapsed at a paper warehouse fire. He was in a coma for 13 years before he died.

The fatalities included 68 volunteer firefighters and 26 career firefighters (down from 33 career in 1994) (Figure 2). Among the volunteer firefighter fatalities, 62 were from local or municipal volunteer fire departments, 3 were part-time or seasonal members of wildland fire agencies, and 3 were members of department fire-police units. All the career firefighters who died were members of local or municipal fire departments. Ninety-one of the fatalities were men and three were women.

The 94 deaths resulted from 89 incidents. Three multi-fatality incidents resulted in 8 firefighter deaths. Four firefighters died in Jackson, Mississippi when a disgruntled firefighter shot coworkers in an administrative meeting. Two firefighters were killed when they lost control of their vehicle, ran off the road, overturned, and hit a tree (the call turned out to be a false alarm). Another two firefighters died in a commercial structure fire when a light weight truss roof collapsed.[2]

Figure 2.
Career vs. Volunteer Deaths

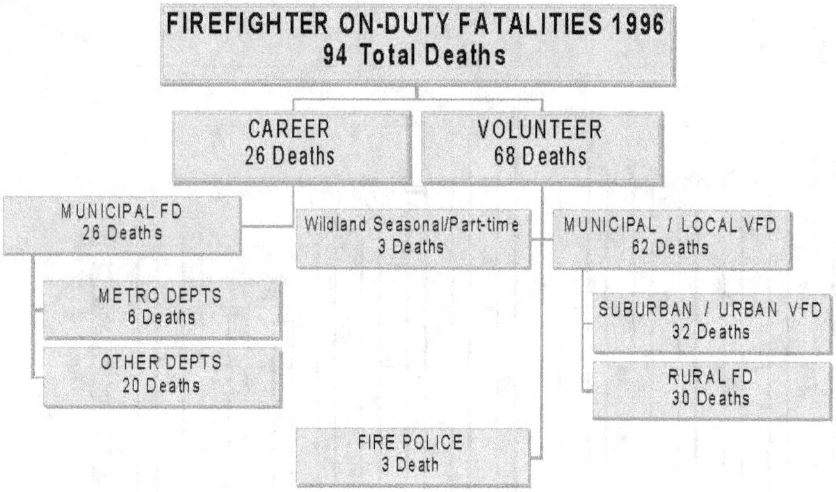

FIREFIGHTER ON-DUTY FATALITIES 1996
94 Total Deaths

CAREER
26 Deaths

VOLUNTEER
68 Deaths

MUNICIPAL FD
26 Deaths

Wildland Seasonal/Part-time
3 Deaths

MUNICIPAL / LOCAL VFD
62 Deaths

METRO DEPTS
6 Deaths

SUBURBAN / URBAN VFD
32 Deaths

OTHER DEPTS
20 Deaths

RURAL FD
30 Deaths

FIRE POLICE
3 Death

The number of deaths associated with brush, grass or wildland firefighting dropped significantly from 18 in 1995 to five deaths in 1996. One firefighter died as a result of dehydration when she became separated from her group on a training run. Four firefighters died of heart attacks, one while repairing a water tender between fires, one after fighting a

---

[2] The report "Two Firefighters Killed in Chesapeake, VA" can be ordered from the USFA.

wildland fire for five hours, one while fighting a wildland fire, and one during a tree-felling class.

Type of Duty

In 1996, 68 firefighter on-duty deaths were associated with emergency incidents, accounting for 72 percent of the 94 fatalities (Figure 3). This includes all firefighters who died while responding to an emergency, while at the emergency scene, or after the emergency incident. Non-emergency activities accounted for 26 fatalities (28 percent). Non-emergency duties include training, administrative activities, or performing other functions that are not related to an emergency incident. One firefighter working at a department fund-raiser and another firefighter directing parade traffic were included in this number.

Figure 3. Firefighter Deaths While Performing Emergency Duty 1996

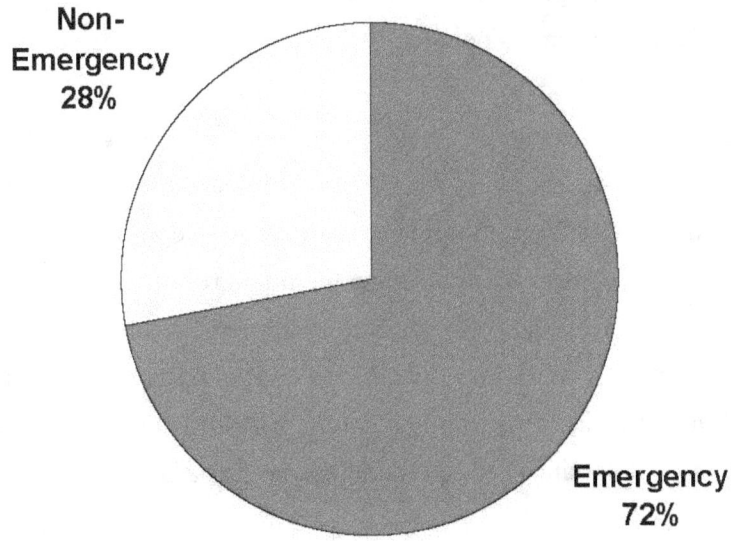

The number of deaths by type of duty being performed is shown in Table 1 and presented graphically in Figure 4. As in previous years, the largest number of deaths occurred during fireground operations. There were 38 fireground deaths, which accounted for 40 percent of the fatalities, down two percent from 1995. Over half (23) resulted from heart attacks on the scene. Eight were from asphyxiation, three from internal trauma, two from electrocution, one from a pulmonary edema, and one from burn injuries.

Table 1. Type of Duty - 1996	Number	Percent
Fireground Operations	38	40.4%
Responding/ Returning from Alarm	22	23.4%
Other/On-Duty	20	21.3%
Non-Fire Emergencies	8	8.5%
Training	6	6.4%
TOTAL	94	100%

The second largest category of deaths by duty type was responding to or returning from emergency incidents, which accounted for 22 deaths in 1996 (down nine deaths from 1995). This has been the second leading cause of deaths since 1993. All 22 deaths involved volunteer firefighters. Six firefighters suffered fatal heart attacks while responding to or returning from emergency incidents. Eight firefighters were killed in fire apparatus accidents while enroute to emergency incidents. At least five of these deaths involved apparatus rollovers. Eight firefighters were killed in accidents involving their personal vehicles while enroute to emergency calls. One of these involved a firefighter who drowned in a roadside lake when his personal vehicle wrecked on the way to a call.

Eight deaths were related to activities at the scene of non-fire emergency incidents. This is down from 13 deaths in 1995. Three firefighters died of heart attacks during EMS incidents. One firefighter was killed when he was electrocuted at a motor vehicle accident. Another firefighter died of asphyxiation during a technical rescue in a grain bin. One firefighter was killed when he was shot by an irate victim at the scene of a motor vehicle accident (MVA), and another firefighter died due to complications from an injury sustained at an MVA that occurred while transporting a patient to the hospital. Another firefighter died when he was hit by a passing motorist while extricating a patient from a vehicle.

Figure 4. Type of Duty 1996

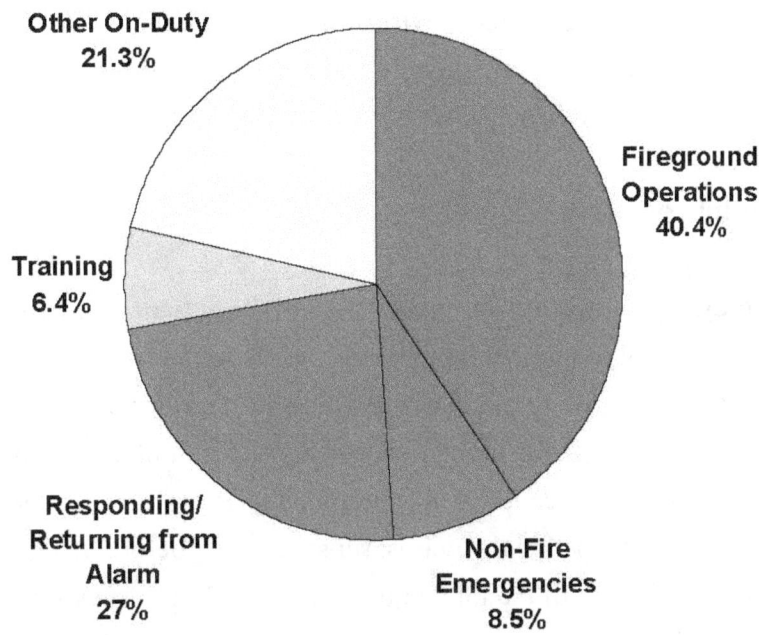

Other On-Duty
21.3%

Fireground
Operations
40.4%

Training
6.4%

Responding/
Returning from
Alarm
27%

Non-Fire
Emergencies
8.5%

There were 26 deaths that occurred during non-emergency duty activities. These deaths include nine firefighters who died from heart attacks while on duty – two at fire department fund-raisers, two during the night at the fire department, two while exercising, one while performing a stress and agility test, one while repairing apparatus between calls, and one while directing traffic for a parade as part of his fire police duties. One firefighter died of a stroke while inspecting fire hydrants. Ten of the 26 non-emergency duty deaths were a result of internal trauma – four firefighters were shot and killed by a disgruntled firefighter during an administrative meeting at the fire department, one firefighter was stabbed to death on the way to storm duty, two firefighters died in motor vehicle accidents (one in transit between station assignments and another while returning from a non-emergency service call), one firefighter was struck by a vehicle while directing traffic, one firefighter fell 20 feet down a fire pole hole in the station, and one was killed during a Fourth of July celebration sponsored by the fire department. The latter was a licensed pyrotechnician who was killed when fireworks prematurely detonated.

Six deaths were attributed to training activities, including one death from dehydration when the firefighter became separated from the group during a training run. Four firefighters died as a result of heart attacks during training – one at a live burn training session, one after an EMS training event, one during a vehicle extrication drill, and one during a wildland tree-felling class. One firefighter died during recruit training from massive organ failure due to a pre-existing condition of sickle cell anemia.

Cause of Fatal Injury

As used in this study, the term *cause* of injury refers to the action, lack of action, or circumstances that directly resulted in the fatal injury, while the term *nature* of injury refers to the medical cause of the fatal injury or illness, often referred to as the physiological cause of death. A fatal injury usually is the result of a chain of events, the first of which is recorded as the cause. For example, if a firefighter is struck by a collapsing wall, becomes trapped in the debris, runs out of air before being rescued, and dies of asphyxiation, the cause of the fatal injury is recorded as "struck by collapsing wall" and the nature of the fatal injury is "asphyxiation". Similarly, if a wildland firefighter is overrun by a fire and dies of burns, the cause of the death would be listed as "caught/trapped," and the nature of death would be "burns". This follows the convention used in NFIRS casualty reports.

Table 2. Cause of Fatal Injury	Number	Percent
Stress or Overexertion	47	50.0%
Struck by or Contact with Object	33	35.1%
Caught or Trapped	7	7.4%
Exposure	5	5.3%
Fell or Jumped	2	2.1%
TOTAL	94	100%

Figure 5 shows the distribution of deaths by cause of fatal injury or illness and Table 2 presents the exact number. As in most previous years, the largest cause category is stress or overexertion, which was listed as the primary factor in 50 percent of the deaths, the same as last year. Firefighting has been shown to be one of the most physically demanding activities that the human body performs. Most firefighter deaths attributed to stress result from heart attacks. Of the 47 stress-related fatalities in 1996, 44 firefighters died of heart attacks, one died of a stroke, one died from complications with sickle cell anemia[3], and one died of a dehydration. Sixteen of the 47 deaths whose cause is listed as stress/exertion occurred during non-emergency activities.

Figure 5. Cause of Fatal Injury

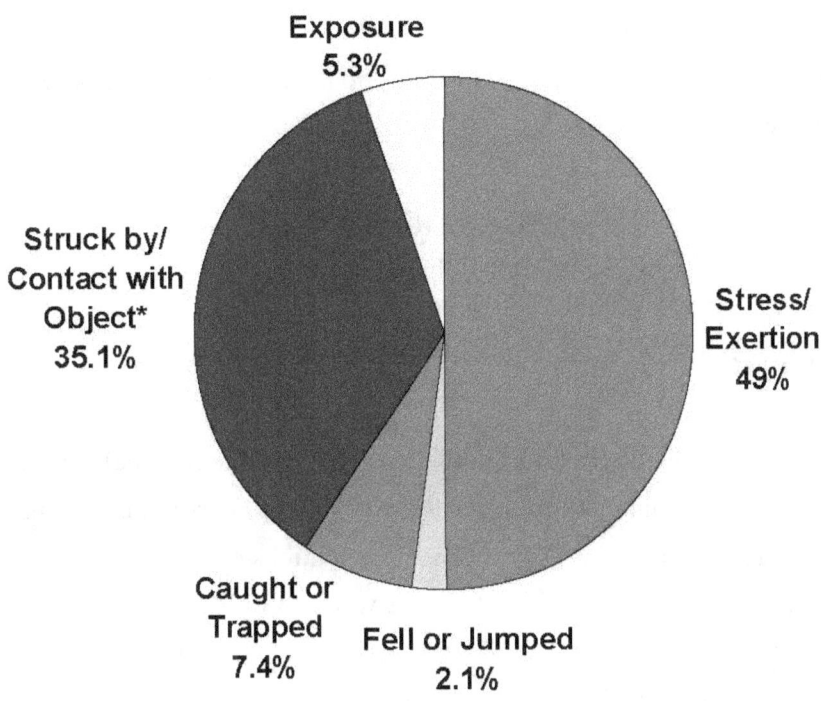

The second leading cause of firefighter fatalities was being struck by or coming in contact with an object. Of the 33 firefighters (35 percent) who died in these incidents, 17 were involved in vehicle accidents, six were murdered, four were struck by vehicles, two were electrocuted, two were killed when their truck was struck by a falling tree, one was struck by detonated fireworks, and one was killed by a collapsing wall.

The third leading cause of firefighter fatalities was being caught or trapped, which accounted for seven deaths (7.4 percent), down 14 percent from 1995. Six firefighters died after becoming trapped by roof collapses – five were in commercial structures and one in a townhouse. One firefighter drowned after being caught in his car after running off the road into a lake.

---

[3] Due to the pre-existing condition of sickle cell anemia, the firefighter after intense exertion went into a state of rahbomyolosis (internal heating and buildup of acid in the heart muscle). This condition caused several major organs to fail resulting in death.

Three asphyxiation deaths were attributed to exposure[4]. One was a firefighter who died when he took off his SCBA (self-contained breathing apparatus) mask in an oxygen deficient atmosphere. One firefighter died from an asthma attack after being exposed to toxic substances at a fire, and another died after inhaling too much smoke while attempting a rescue at a structure fire. Two other firefighters died from exposure. One died as a result of burns after he was caught in a flashover, and the other died from cardiac arrest/pulmonary edema after being exposed to a cloud of unknown chemical vapors at a commercial structure fire.

Two firefighters died as a result of falls. One firefighter slipped and fell down a fire pole about 20 feet. Another firefighter died after falling and striking his head.

Nature of Fatal Injury

Table 3 and Figure 6 show the distribution of the 94 deaths by the medical nature of the fatal injury or illness. The leading nature of death in 1996 was heart attacks, which accounted for 46 firefighter fatalities. Two of the heart attacks occurred while exercising, and one occurred during an agility/stress test. There were 24 firefighters who suffered heart attacks at fire scenes[5] and five who suffered heart attacks enroute to or returning from calls. Three heart attacks occurred at EMS or rescue incidents. Three others occurred during fund-raisers, two while asleep at the fire station, one while repairing fire apparatus, one just following a fire, and four during training (Figure 6a).

Table 3. Nature of Fatal Injury	Number	Percent
Heart Attacks	46	48.9%
Internal Trauma	31	33.0%
Asphyxiation (includes drowning)	10	10.6%
Electrocution	3	3.2%
Dehydration	1	1.1%
Stroke/Seizure	1	1.1%
Burns	1	1.1%
Other (sickle cell anemia)	1	1.1%
TOTAL	94	100%

---

[4] "Exposure/Contact with" follows NFIRS 4.0 definitions under "Cause of Fatal Injury".
[5] One firefighter included in this total died of cardiac arrest as a result of an exposure at a hazmat incident.

Internal trauma was the second leading nature of death, responsible for 32 deaths (up nine from 1995). This total includes 18 firefighters who were involved in vehicle accidents, four who were hit by vehicles while on the emergency scene, and six firefighters who were victims of violence. Four other firefighters died as a result of internal trauma – one firefighter died after falling down a fire pole hole, one died when fireworks detonated at a Fourth of July celebration, one was crushed by a collapsing wall, and one was killed when a tree fell on the fire apparatus.

Figure 6. Nature of Fatal Injury

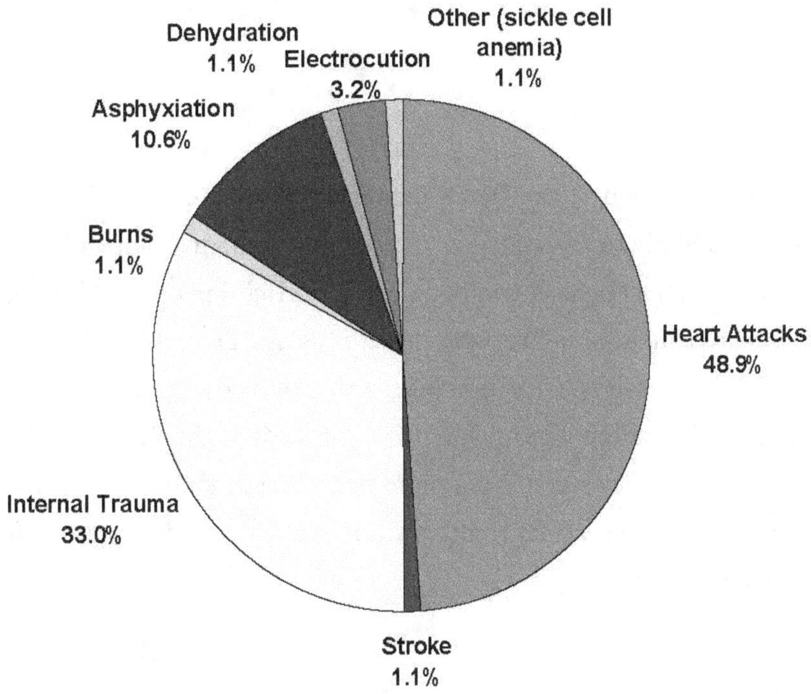

Dehydration 1.1%
Electrocution 3.2%
Other (sickle cell anemia) 1.1%
Asphyxiation 10.6%
Burns 1.1%
Heart Attacks 48.9%
Internal Trauma 33.0%
Stroke 1.1%

Asphyxiation was the third leading medical reason for firefighter deaths, responsible for 10 deaths (50% less than 1995). Of the ten firefighter deaths, eight resulted from carbon monoxide poisoning or inhalation of smoke or superheated gases during structural firefighting. All of these eight deaths occurred when the firefighters became caught and trapped by rapidly spreading fires or structural collapses. Two other firefighters died as a result of asphyxiation – one when he took his SCBA mask off in an oxygen deficient atmosphere and another when he became trapped and drowned after wrecking his car in a lake.

17

Three firefighters died from electrocution – one whose SCBA became caught in downed power lines, one who came into contact with downed power lines at an automobile accident, and one who had power lines fall on him when a power pole broke and fell to the ground.

Only one of the 94 firefighter fatalities was attributed to burns. The firefighter was caught in a flashover in a two-alarm fire in an apartment.

The medical causes of death for the final three were dehydration, stroke, and organ failure due to a pre-existing condition of sickle cell anemia.

Firefighters Ages

Figure 7 shows the distribution of firefighter deaths by age and cause of death. Younger firefighters were more likely to have died as a result of traumatic injuries from an apparatus accident or after becoming caught or trapped during firefighting operations. Stress was shown to play an increasing role in firefighter deaths as age increased. This is also reflected in Figure 8, which shows the distribution of deaths by age and medical nature of injury. Trauma and asphyxiation were responsible for most of the deaths of younger firefighters, while heart attacks were much more prevalent among older firefighters. Heart attacks accounted for 16 of the 26 firefighters who were over 50 years old, and all 12 of the firefighters over 60 years old.

Figure 7. Age & Cause*

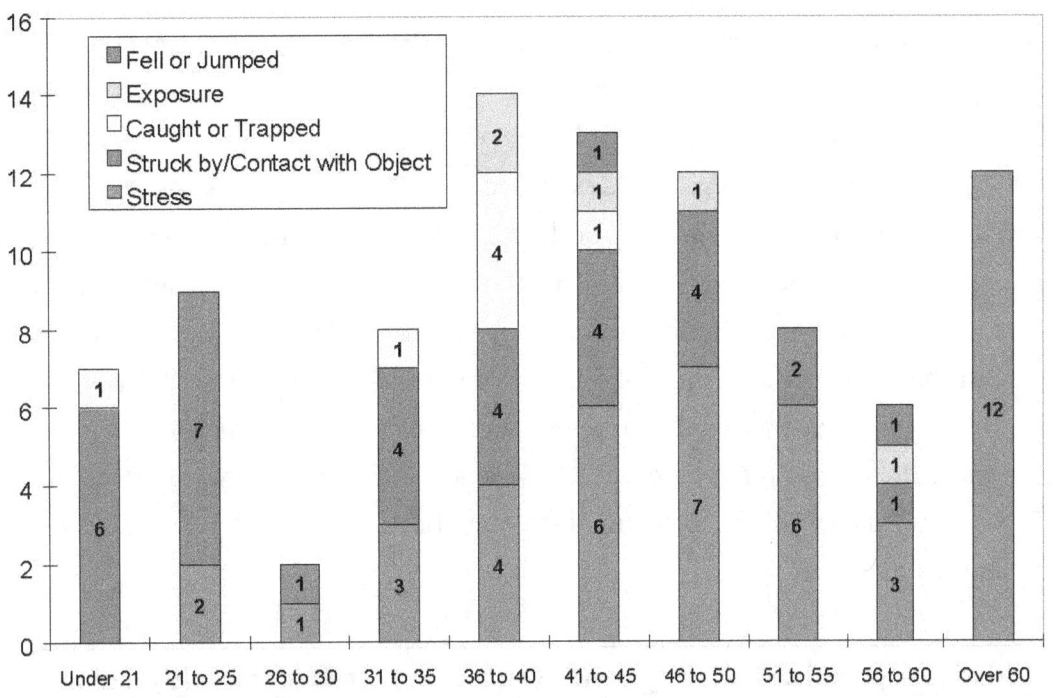

Figure 8. Age & Nature*

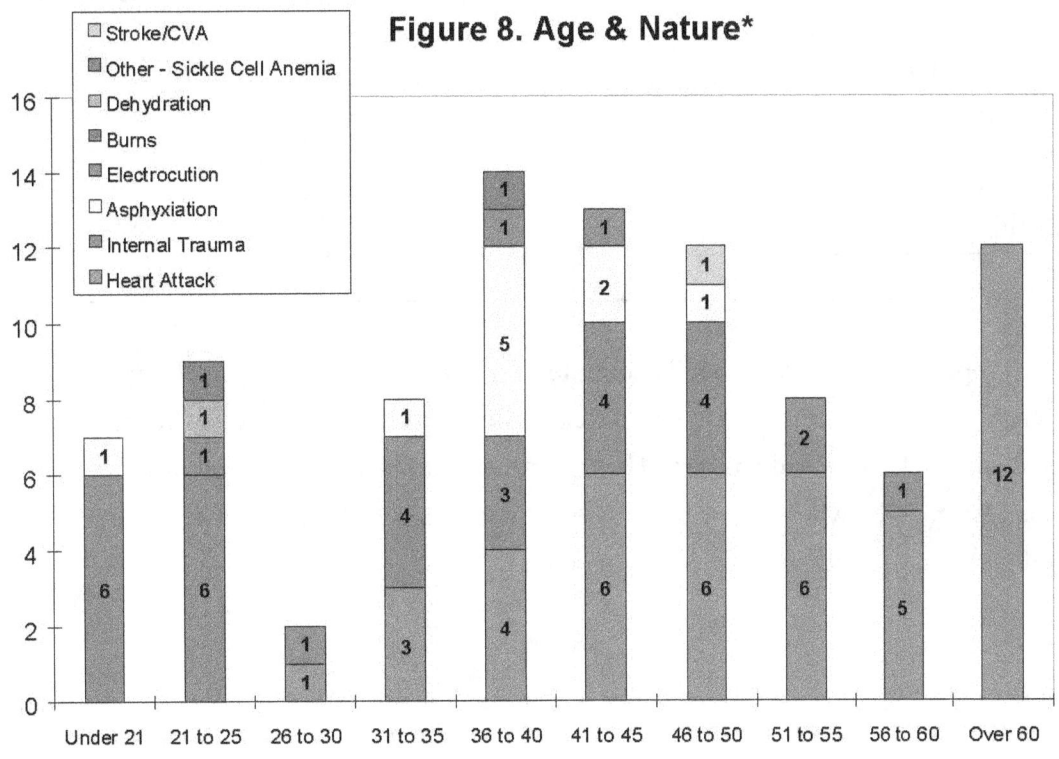

Fireground Deaths

There were 38 fireground deaths in 1996, a decrease of two from 1995. Figure 9 and Table 4 show the distribution by fixed property use.

Property Type – 31 of the 38 fireground deaths occurred at structure fires. As in most years, residential occupancies accounted for the highest number of these fireground fatalities, with 19 deaths (50 percent). Residential occupancies usually account for 70-80 percent of all structure fires and a similar percentage of the civilian fire deaths each year, 50 percent of the firefighter deaths in 1996 occurred in residential structures.[6] The frequency of firefighter deaths in relation to the number of fires is much higher for non-residential structures. One firefighter died in 1996 in a storage occupancy compared to six in 1995. Nine firefighters died in commercial structure fires, one in public assembly, and one in manufacturing.

Table 4. Fixed Property Use for Fireground Deaths	Number	Percent
Residential	19	50.0%
Commercial	9	23.7%
Outdoor Property	5	13.2%
Street/Road	2	5.3%
Storage	1	2.6%
Public Assembly	1	2.6%
Manufacturing	1	2.6%
TOTAL	38	100%

Outdoor properties and "street/road" accounted for a total of seven deaths. Four out the five outside property deaths were heart attacks. The fifth was a firefighter who was pinned between two trucks at a wildland fire. The two street/road deaths consisted of a firefighter who was electrocuted when a power line came down at a pole fire, and one who had a heart attack while directing traffic at an emergency scene.

---

[6] Complete NFIRS data for 1996 fire incidence was not available at the time of this report, but residential fires typically account for between 70 and 80 percent of all civilian fatalities each year.

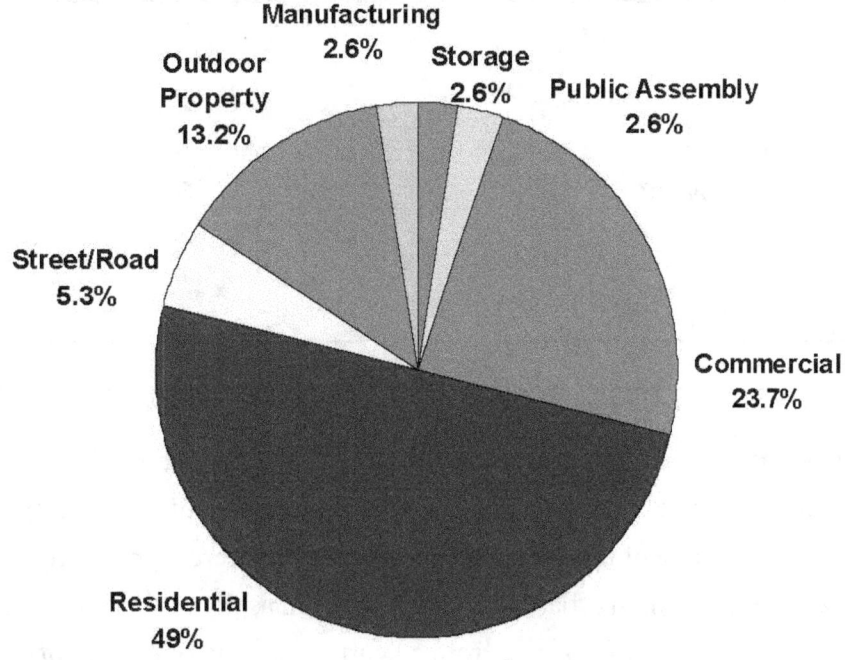

Figure 9. Fixed Property Type

- Manufacturing 2.6%
- Outdoor Property 13.2%
- Storage 2.6%
- Public Assembly 2.6%
- Street/Road 5.3%
- Commercial 23.7%
- Residential 49%

Type of Activity

Figure 10 and Table 5 show the activities the 38 firefighters were engaged in at the time they sustained their fatal fireground injuries or illnesses. There was a substantial decrease this year compared to last year in the number of firefighters who died while engaged in traditional engine company duties of fire attack and advancing hose lines (decrease of 14). Nine firefighters died while performing these fireground operations, including six who died from asphyxiation after becoming trapped by rapid fire spread or structural collapse while advancing hose lines. Three other firefighters suffered heart attacks while performing similar functions. Eight firefighters died while performing water supply operations on the fireground – five from heart attacks, one from electrocution, one from being pinned between two fire apparatus, and one from being struck by a passing motorist.

Table 5. Type of Activity for Fireground Deaths	Number	Percent
Advancing Hose Lines/ Fire Attack	9	23.7%
Support Duties	9	23.7%
Water Supply	8	21.1%
Search and Rescue	5	13.2%
Cutting Fire Breaks (Wildland)	3	7.9%
Ventilation	2	5.3%
Incident Command	1	2.6%
Salvage & Overhaul	1	2.6%
TOTAL	38	100%

Traditional truck and ladder company duties accounted for eight deaths. Search and rescue operations in burning structures were being conducted when five of these deaths occurred, no change from 1995. Three of the search and rescue deaths were from heart attacks, one from burns after being caught in a flashover, and one from asphyxiation. Two firefighters died while ventilating structure fires, one from a heart attack and one from internal trauma when a wall collapsed. One firefighter was electrocuted during salvage and overhaul operations when a power line came into contact with his SCBA.

Nine firefighters died while performing support functions or standing by on the fireground – seven from heart attacks, one from asphyxiation, and one from internal trauma.

Cutting fire lines to contain grass, brush, and forest fires accounted for three firefighter fatalities. All three died as a result of heart attacks.

One incident commander suffered a fatal cardiac arrest/pulmonary edema at fire incidents.

Figure 10. Type of Activity

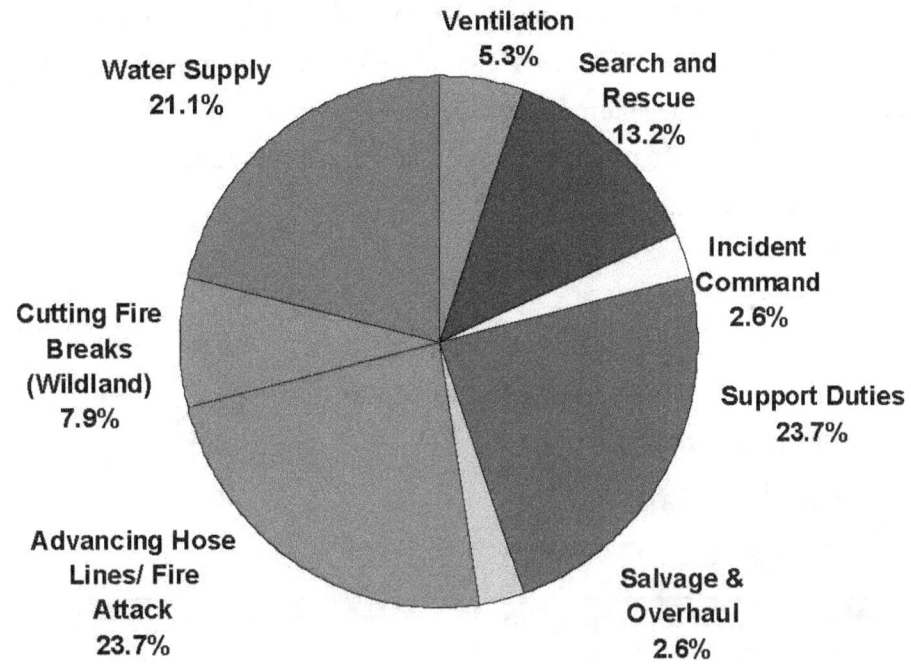

Ventilation
5.3%

Water Supply
21.1%

Search and
Rescue
13.2%

Incident
Command
2.6%

Cutting Fire
Breaks
(Wildland)
7.9%

Support Duties
23.7%

Advancing Hose
Lines/ Fire
Attack
23.7%

Salvage &
Overhaul
2.6%

Time of Alarm

The distribution of 1996 deaths according to the time of day when the incidents were reported is shown in Figure 11 (49 times were not reported). The highest number of fireground deaths occurred for alarms that were received between 2300 and 0059. The second highest number was a tie between 1700-1859 and 1100-1259. There were no fireground deaths between the hours of 0700-0859.

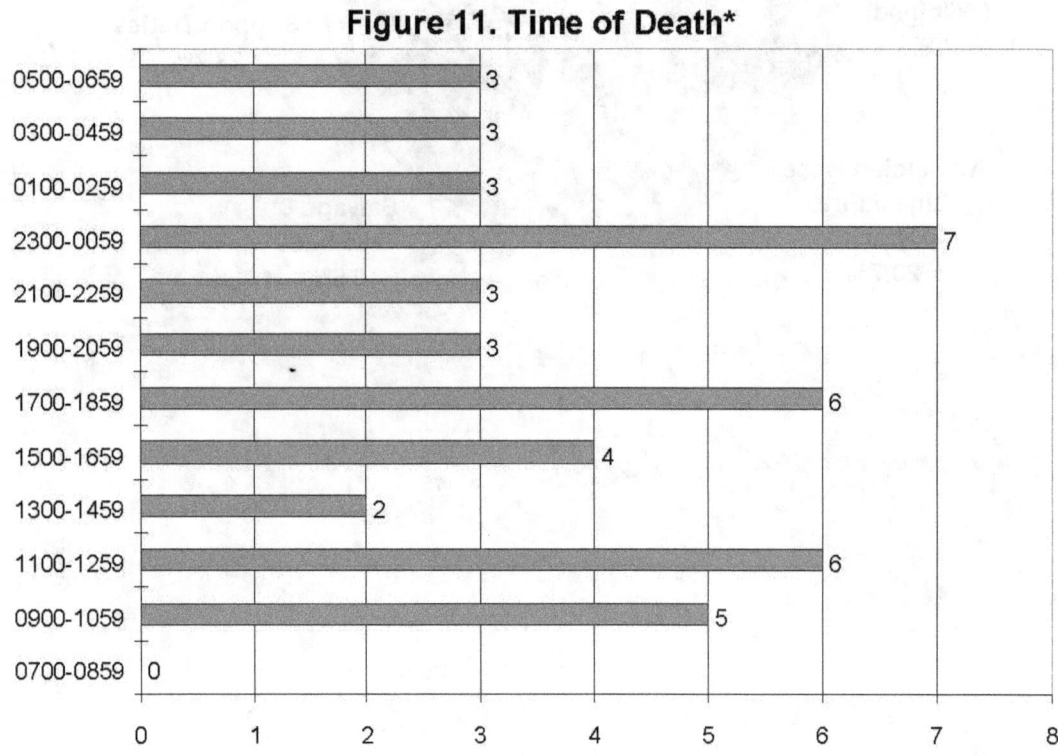

Figure 11. Time of Death*

Month of the Year

Figure 12 illustrates firefighter fatalities by month of the year. Firefighter fatalities peaked in January and April. Other high months were recorded in August and October. The early summer months (May, June, and July) were among the lowest months. (Conversely the number of residential fires peaked during the winter and was lowest during June and July.)

Figure 12. Deaths by Month of the Year

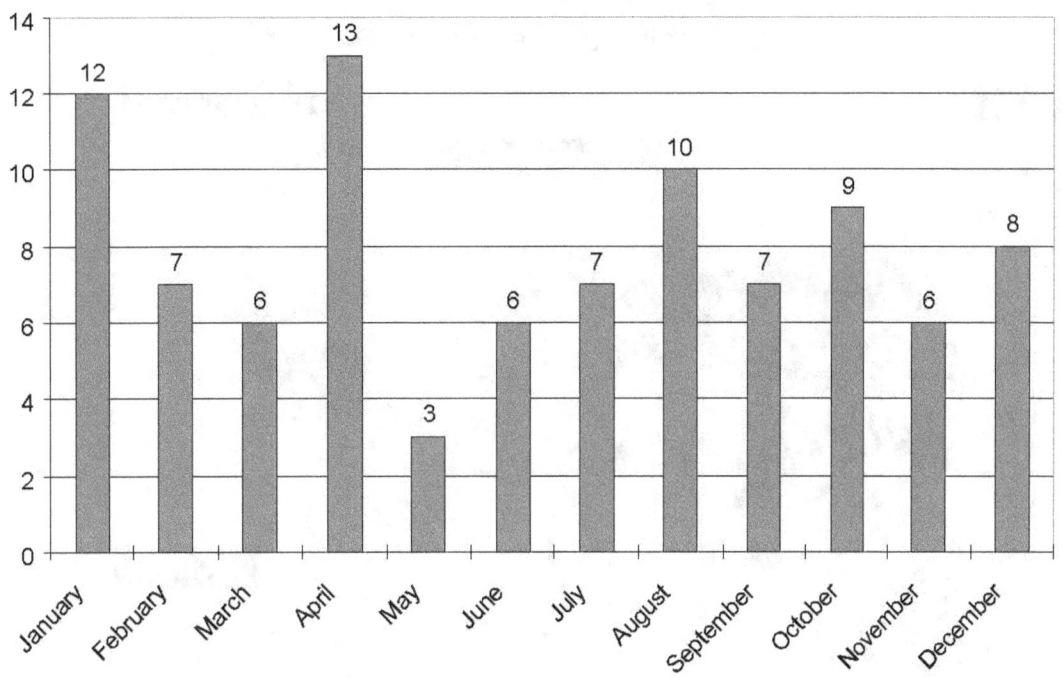

State and Region

The distribution of firefighter deaths by state is shown in Table 6.[7] Thirty-four states had at least one firefighter fatality. New York led with 13 deaths. Figure 13 shows the firefighter fatalities divided by region of the country and whether they were career structural, volunteer structural, or career or seasonal wildland firefighters.

Figure 13.
Firefighter Deaths By Region 1996

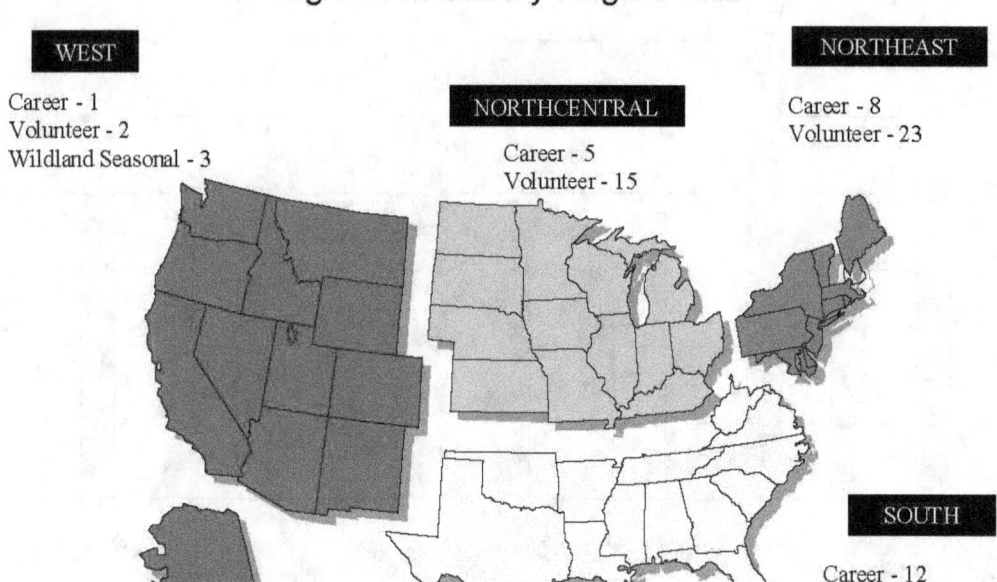

WEST
Career - 1
Volunteer - 2
Wildland Seasonal - 3

NORTHCENTRAL
Career - 5
Volunteer - 15

NORTHEAST
Career - 8
Volunteer - 23

SOUTH
Career - 12
Volunteer - 25

---

[7] This list attributes the deaths according to the state where the fire department or unit is based, as opposed to the state where the death occurred. They are listed by those states for statistical purposes, and for the National Fallen Firefighter Memorial at the National Fire Academy.

Table 6.
1996 State with On-Duty Firefighter Fatalities

State	Number of Deaths	State	Number of Deaths
Alabama	2	Nebraska	2
Arkansas	1	Nevada	1
Arizona	1	New Jersey	5
Connecticut	2	New York	9
Georgia	4	North Carolina	3
Hawaii	2	Ohio	3
Illinois	4	Oklahoma	4
Indiana	4	Pennsylvania	6
Iowa	1	South Carolina	4
Kansas	1	Tennessee	1
Kentucky	1	Texas	5
Louisiana	2	Utah	1
Maine	1	Vermont	1
Maryland	5	Virginia	3
Massachusetts	2	Washington	1
Michigan	2	West Virginia	2
Mississippi	4	Wisconsin	2
Missouri	1	Wyoming	1

Total: 94

Analysis of Urban/Rural/Suburban Patterns in Firefighter Fatalities

The US Bureau of the Census defines "urban" as a place having a population of at least 2,500 or lying within a designated urban area. Rural is defined as any community that is not urban. Suburban is not a census term but may be taken to refer to any place, urban or rural, that lies within a metropolitan area defined by the Census Bureau, but not within one of the central cities of that metropolitan area.

Fire department areas of responsibility do not always conform to the boundaries used for the census. For example, fire departments organized by counties or special fire protection districts may have both urban and rural coverage areas. In such cases, it may not be possible to characterize the entire coverage area of the fire department as rural or urban, and firefighter deaths were listed as urban or rural based on the particular community or location in which the fatality occurred.

The following patterns were found for 1996 firefighter fatalities. These are estimates based upon population and area served reported by the fire departments.

Table 7.

	Urban/Suburban	Rural	Federal or State Parks/Wildland	Total
Firefighter Deaths	61	30	3	94

SPECIAL TOPICS

Homicides and Violence in the Workplace

Violence towards emergency services providers is a growing concern. In 1996, six firefighters were murdered while on duty, including four in a single shooting incident. In numerous other cases, firefighters have been the victims of violence perpetrated by random attackers, by the citizens they have attempted to serve, and by fellow workers. The fire service is no longer immune to these events and trends which have been affecting other segments of society for many years.

National statistics on workplace violence are a cause for increasing alarm. Homicide has become the third leading cause of death in the workplace. Murders in the workplace are also the fastest growing type of murder in the United States. The murder rate of supervisors has doubled since 1985. Statistical information shows that the murder rate for public sector employees is more than twice that of their private sector counterparts.[8]

The fire service does not adequately track violence directed towards employees and volunteers. Reports of violence vary widely by state and locality. No NFPA standard deals directly with the issue of reducing violence towards emergency workers, except the portion of NFPA 1500, Standard on Fire Department Occupational Safety and Health Program, that references response to civil disturbances. The Occupational Safety and Health Administration (OSHA) has a policy that employers must provide a workplace free of violence, however applying this policy to firefighters is difficult, because they, like paramedics and police officers, work on the streets in good and bad neighborhoods, under stressful situations, and without the safety that can be engineered into an individual worksite.

The six murdered firefighters represent 6.4 percent of firefighter deaths in 1996. This contrasts with the traumatic (includes asphyxiation) deaths of only seven firefighters while engaged in interior firefighting during the same year.

---

[8] Barrett, Stephen. "Protecting Against Workplace Violence," *Public Management*, August 1997. Pages 9-12.

Six Murders - In the most highly publicized workplace violence incident in 1996, six Jackson (MS) Fire Department career fire officers were shot by a fellow firefighter while attending a staff meeting on April 24. Four of the officers were wounded fatally. The firefighter murdered his estranged wife earlier in the day, and went to the city's central fire station, where he barged into the staff meeting and started shooting. He also tried to force his way into the Fire Chief's office. He was apprehended by police after a chase and shoot-out.

Fellow firefighters indicated that the individual had previously expressed his hatred of chief officers. Widespread racial tensions in the department are also thought to have contributed to the motivation for the shooting incident. Police later linked one of the weapons used by the firefighter to two previously unsolved murders in Jackson.

In an incident that occurred on January 7, 1996 in Pleasantville, New York, a firefighter was murdered while enroute to his station. The volunteer firefighter was walking to the station during a blizzard to report for storm duty, but he never arrived. His body was discovered several hours later. His throat had been slashed and he had been stabbed and bludgeoned to death. The police have not made any arrests in this case.

In the third incident, an 18 year old volunteer firefighter was shot to death in Indiana when he was the first to arrive at the scene of a motor vehicle accident involving a car and a motorcycle. The driver of the car shot the motorcyclist and the firefighter, killing them both.

Preventative Strategies – OSHA separates workplace violence into three categories based upon the perpetrators' links to the workplace. Type I violence occurs when an employee is the victim of a random act of violence from an offender with no apparent ties to the workplace. This could have been the case in the murder of the firefighter in New York. Type II violence occurs when the attacker is a recipient of the victim's services, such as a patient attacking an EMS crew. This was the case in Indiana. Type III violence occurs when the attacker has a direct relation to the workplace, such as a fellow employee, former employee, or relative or acquaintance of an employee. This was the case in Jackson, Mississippi.

Firefighters are at increased risk for all three types of violence. Stations, with few exceptions, are open buildings widely visited by the public. The nature of firefighting, rescue, and EMS work brings the employee or volunteer into very dangerous areas of their communities, often directly into the path of domestic feuds or other violent situations. Customers may be mentally ill or under the influence of drugs or alcohol, making their actions erratic, unpredictable, and violent. Compounding these external dangers are the pressures felt by tired and stressed co-workers due to work or personal problems. Risks are increased when families or acquaintances become involved.

There are strategies that departments can take to reduce violence or death to workers. Random violence may be reduced through improvements in station design and security. Firefighters should be made aware that they can be victims while on-duty, and that they may be targeted because of their job and profile in the community, or because they work at all hours in all types of areas. Violence towards workers by patients and bystanders at incident scenes may be reduced by training firefighters in verbal de-escalation methods, self-defense tactics, and survival training, similar to the training that police officers receive. They should be trained to recognize and avoid a dangerous situation or conflict in the first place, to try to verbally defuse it if they become involved, and to defend themselves to survive or escape an attack. Additionally, police should escort units to high crime areas and certain types of incidents. Body armor should be made available to any firefighters who respond to high crime areas. Firefighters and EMS workers should be clearly identified as different from police officers. Departments should consider distinct uniforms, ones that clearly identify personnel as firefighters when they are not in their turnout gear, but do not look like local police uniforms.

In the case of type III violence by coworkers, departments across the United States should consider several improvements in their recruitment and evaluation processes for employees or volunteers. Few departments do psychological profiling of members in addition to thorough background checks. This should be expanded, to help identify applicants who may not be suited for the particular stresses of a firefighting career. Employees should be monitored throughout their careers for emotional and mental fitness and stability. Stress reduction and wellness programs should be expanded to help prevent firefighters from succumbing to mental stressors. Employee assistance programs and mental health professionals should be available on a confidential basis for employees who need psychological assistance. Conflict management training and sensitivity training should be provided at all levels, not just for managers. Confidential reporting systems should be developed to help overcome the culture of non-reporting of fellow workers who become a threat to themselves or to others. Mechanisms for psychologically assisting discharged employees through their transition into other jobs should be considered.

Firefighting is dangerous enough without the additional stress of threatened violence. By focusing on training firefighters to operate more safely in dangerous situations, by properly screening prospective employees, and by providing continual psychological counseling for firefighters and their families, the fire service can hopefully prevent some of these tragic deaths in the future.

Firefighter Health and Wellness

1996 saw a continued positive emphasis on firefighter health and wellness throughout the fire service. A highlight in this area was the cooperative effort to develop firefighter physical fitness and wellness programs agreed to by the International Association of Fire Fighters (IAFF) and the International Association of Fire Chiefs (IAFC). In concert with 10 fire departments throughout the United States and Canada, the IAFF and IAFC will jointly design a program that can be used by fire departments everywhere to improve the health and wellness of firefighters.

The development and implementation of health and wellness programs for firefighters is instrumental for reducing the annual number of firefighter injuries and fatalities. Under ideal conditions, firefighting is a strenuous task requiring an above-average level of physical fitness. Additional demands are placed on firefighters by the many inherent stresses of the job, which can affect their physical and emotional health and well being. The nature of firefighters' unusual work and eating schedules also requires that attention be paid to wellness efforts like dietary education and modification, body composition testing, and stress reduction.

Data collected for 1996 indicates that improvements in firefighter health and wellness are still sorely needed: 46 fatalities occurred in 1996 as a result of heart attacks, of which 45 were attributed to stress or overexertion, with one attributable to cardiac arrest subsequent to a hazardous materials exposure. Heart attacks were the leading cause of firefighter deaths, representing 48.9% of all 1996 fatalities.

Heart attacks struck career (13), volunteer (31), and wildland seasonal (2) firefighters with ages ranging from 29 to 78. These firefighters were performing a variety of activities including: advancing hose lines, search and rescue, ventilation, fireground support, training, water supply, cutting fire breaks, and station duties, among others.

The 1996 data clearly suggests that there is still a great deal of work to be done in the firefighter health and wellness arena. Both career and volunteer firefighters stand to benefit from the cooperative effort currently being undertaken by the IAFF and IAFC. Fire departments will also benefit from the reduced casualty rates that can be expected with improved firefighter health and wellness.

Vehicle Accidents

In recent years, a great deal of emphasis has been placed on the safe operation of fire department vehicles while responding to, or returning from, incidents. Emergency vehicle operator courses (EVOC) and defensive driving classes are increasingly being required for drivers of fire department vehicles. Commercial driver's licenses (CDLs) or special vehicle operator's endorsements may also be required by individual states, although many states still exempt drivers of fire department apparatus from these requirements. Some states offer comprehensive driver/operator courses in a modular format that allows them to be presented to both career and volunteer personnel, although the completion of such a course may or may not be required for vehicle operators. Several fire departments have implemented response procedures designed to minimize the amount of emergency or "lights and siren" driving required, by prioritizing response levels such that later-arriving units respond in a routine or "non-lights and siren" mode.

Despite the recent emphasis on vehicle safety throughout the fire service, the 1996 data indicates that more work needs to be done in this area, as 22 firefighter deaths occurred as a result of vehicle accidents. Although the number of deaths from vehicle accidents is less than last year (down 8 deaths from 1995) they still represent 22.4% of all 1996 firefighter fatalities. For 1996, as in every year since 1993, vehicle accidents remain the second leading cause of firefighter fatalities.

The causes of the enroute fatalities were varied[9]: eight firefighters were killed as a result of trauma incurred during fire apparatus accidents, five of those involved apparatus rollovers. Eight firefighters were killed while operating their personal vehicles enroute to emergency calls, including one who drowned in a lake after his vehicle wrecked.

All 16 of the firefighters killed as a result of vehicle accidents were volunteers.[10] This data may indicate that further efforts are needed to train and qualify volunteer firefighters as emergency vehicle operators. The nature of the volunteer service may make it difficult for vehicle operators to maintain proficiency or familiarity with the unique driving

---

[9] Five firefighters that suffered from fatal heart attacks while responding to or from incidents are not mentioned in this section.

[10] Five firefighters that suffered from fatal heart attacks while responding to or from incidents are not mentioned in this section

34

requirements of fire apparatus. Defensive driving and vehicle operator training is also important for volunteers who respond to emergencies in their personal vehicles. Such training may have the added benefit of reduced insurance premiums for members and the fire department.

In order to reduce the number of firefighter fatalities related to vehicle accidents, firefighters must remember that emergencies will only worsen unless the fire department response is conducted with a high degree of safety, ensuring that personnel and equipment arrive unharmed and ready to do the job. The development of safe firefighter attitudes coupled with improved training and certification requirements should help reduce the number of vehicle accidents that occur while responding to, or returning from, incidents. Reducing the number of fire department vehicle accidents should contribute to a reduction in the number of firefighter casualties resulting from these accidents.

CONCLUSIONS

The analysis of firefighter deaths in 1996 indicates that the overall long-term trend toward fewer firefighter fatalities is continuing. The 94 fatalities in 1996 are the third lowest recorded, and only the third time the total number of fatalities has dropped below 100, all within the last four years.

Stress-induced heart attacks continue to remain the number one cause of firefighter deaths.[11] Health and fitness programs should help to reduce these numbers in the long term. Better screening for high risk firefighters through medical exams may help to prevent some deaths in the short term, by identifying firefighters who are unfit for strenuous duty or who may be at high risk for heart disease. Many factors that place firefighters at high risk for heart attacks are controllable, such as better nutrition, not smoking, and exercise.

Figure 14
Stress vs. Actions

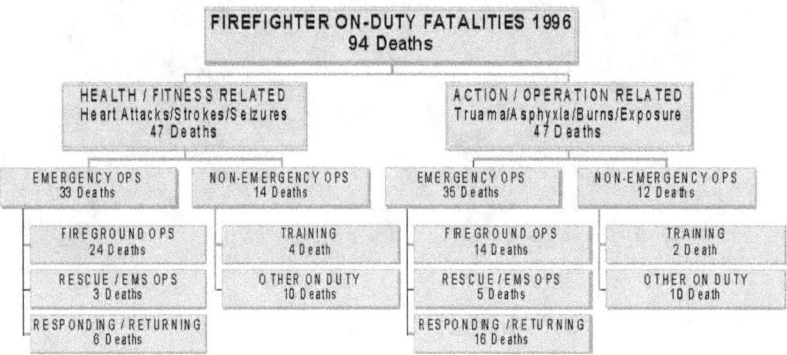

Six firefighters died of asphyxiation in structure fires, all the result of a structural collapse. These and several other incidents reinforce the need for proper size-up, continual progress reports, and a working accountability system at all incidents to keep track of personnel. An accountability system should track all members, including who and where they are, who they are working for, what they are doing, and how long they have been doing it.

---

[11] Figure 14 shows the relation between stress-related and action-related firefighter deaths.

Perhaps the best way to reduce fireground fatalities is through the adoption of more stringent building codes, better fire prevention, and strict code enforcement.

The institution of rapid intervention teams into emergency operations may help save more firefighters' lives. PASS devices should be used at all fire incidents, to improve the chances of being alerted about, and locating, a downed firefighter.

Responding to incidents continues to claim too many lives. In 1994, personal vehicle accidents as a cause of fatalities were almost eliminated, however, in 1995 the number of vehicle accident deaths rose and continued into 1996. All fire departments should have a policy regarding driver training for responding to emergency incidents, and all drivers should use caution when approaching intersections. Also, seat belts can only save lives if they are worn; they were not used in several 1996 fatal accidents.

As in last year's analysis, the 1996 statistics indicate that risk management, in the form of assessing firefighters' health risks, sizing up fireground conditions, and evaluating hazards at special rescue scenes, is a key to reducing on-duty firefighter fatalities even further.

Firefighting is a dangerous occupation and the additional stress of threatened violence is not needed. Fire departments, like many other industries are doing, should develop training to teach firefighters to operate more safely in dangerous situations, properly screen prospective employees, and provide continual psychological counseling for firefighters and their families. As a result, the fire service can hopefully prevent some of these tragic deaths in the future.

APPENDIX A

1996 INCIDENT SUMMARY

1996 INCIDENT SUMMARY

Incident 1

FDNY, New York, NY
On January 5th, James B. Williams, a career firefighter with FDNY, died from burns sustained during a 2-alarm fire rescue at an apartment building in Queens, New York. Unaware that the occupants of the apartment had already left, he and four other firefighters were searching for victims and fighting the fire when they were engulfed in flames after breaking through a door.

Incident 2

West End Fire Company, Stowe, Pennsylvania
On January 5th, William R. Favinger, Sr., a volunteer for the West End Fire Company, suffered a fatal heart attack while returning from an automatic alarm. He collapsed in the station while filling out the roster. CPR was immediately started and paramedics were called.

Incident 3

Owego Fire Department, Newark, New York
On January 6th, volunteer firefighter Guy R. Polllard, suffered a fatal heart attack while performing pump operations at a suspected house fire on a mutual call with the Owego Fire Department. After determining that the house's chimney was stuffed, trucks begin preparing to leave when Firefighter Pollard suffered the heart attack. He was transported to Wilson Memorial Regional Medical Center where he was pronounced dead on arrival.

Incident 4

Pleasantville Fire Department, Pleasantville, New York
On January 7th, Firefighter Thomas Dorr, was killed as a result of multiple injuries (stab wounds) while responding to the station on foot during a snow storm. He was walking to the firehouse for storm duty and was attacked at some point.

Incident 5

Rockaway Borough Fire Department, Rockaway, New Jersey
On January 7th, volunteer Firefighter Willard Hopler, suffered a massive MI while operating an aerial apparatus at the scene of a chimney fire. Despite resuscitative efforts that were performed by his crew, he was pronounced dead at Northwest Covenant Medical Center.

Incident 6

Pecatonica Fire Protection District, Pecatonica, Illinois
On January 13th, Chief Dale Zimmerman, died in an attempt to rescue two men who were overcome by fumes in a grain bin. During the rescue, alarm bells went off warning that another firefighter's air tank was getting low. The Chief went over to change the tank, but his mask was fogging up. He took off his mask in order to change the tank. The other fireman then left, but saw the chief having problems in his attempt to rescue the two men. The two men were rescued, but the chief eventually died from asphyxia from the carbon monoxide poisoning.

Incident 7

Cairo Fire Department, Cairo, Georgia
On January 18th, Firefighter Marcel Glenn, died while fighting a structure fire. Firefighter Glenn was ventilating a house fire by breaking the windows with a fire hose. After he had ventilated two windows, he turned around and collapsed. EMS was called and he was taken to the hospital where he was pronounced dead due to cardiac arrest.

Incident 8

Citizens Hose Company #1, South Renova, Pennsylvania
On January 19th, volunteer Firefighter Reed Morton Sr., suffered a heart attack while directing traffic and assisting evacuees at a fire (fire policeman).

Incident 9

Wildwood Fire Association, Alger, Michigan
On January 19th, volunteer Firefighter Robert Haggadone, was struck by a passing motorist while working a hydrant at a house fire. He died after being in a coma for 7 1/2 months.

Incident 10

Dallas Fire Department, Dallas, Texas
On January 22nd, recruit Firefighter Jerald Dibbles, died during his second day at the training academy. He had a pre-existing condition of sickle cell anemia (trait). Because of this condition he went into a state of rahbomyolosis (internal heating and buildup of acid in the heart muscle). This condition caused several major organs to fail resulting in death.

Incident 11

Livingston Fire Department, Livingston, Texas
On January 26th, volunteer Firefighter Dale Burkhalter, died in a car accident while returning from a fire incident. The accident occurred at four in the morning at a dark and unlighted intersection. The road conditions were wet and there were patches of fog. In an attempt to cross a major highway, the firefighter's car was struck on the driver's side.

Incident 12

Lake Murray Village Fire Department, Ardmore, Oklahoma
On January 31st, volunteer Marvin Maphes, drove a tanker to the fire scene and suffered a fatal heart attack on arrival.

Incident 13

Kauai County Fire Department, Kauai, Hawaii
On February 1st, Firefighter Steven Gushiken, got up in the morning (4:30 a.m.) while still on-duty and went for a walk in the park adjacent to the station (normal morning routine). When the rest of the shift woke up later they found him unconscious on the ground. Coworkers unsuccessfully attempted to revive him.

Incident 14

FDNY, Brooklyn, New York
On February 5th, Firefighter Louis Valentino became trapped when the roof of an Auto Body Shop in East Flatbush-Brooklyn collapsed. Fifteen other firefighters were injured at this 3-alarm blaze. Firefighter Valentino died less than an hour after the fire started at about 3:40 p.m.

Incident 15

Clarksville Fire Department, Clarksville, Virginia
On February 5th, Firefighter Corey Morgan, died in a motor vehicle accident while responding to a fire call.

Incident 16

Ridgefield Boro Fire Department, Ridgefield, New Jerse
On February 11th, Firefighter Michael McLaughlin, died of an apparent heart attack after arriving on the scene of a small fire in a laundromat. Firefighter McLaughlin experienced head trauma when he fell on the scene and knocked his head against the fire engine. This trauma in turn resulted in cardiac arrest.

Incident 17

Enville Fire Department, Marietta, Oklahoma
On February 11th, volunteer Firefighter Raymond Vinson, fought a grass fire for approximately seven hours in the morning, when he was called out again for another grass fire. This incident lasted about five hours. He died of a heart attack after returning from the incident.

Incident 18

I.X.L. Fire Department, Castle, Oklahoma
On February 23rd, Firefighter Nathaniel Quinn went into cardiac arrest while fighting a wildland fire near Okemah. The wildfires blackened up to 30,000 acres, and destroyed 43 homes in 10 Oklahoma counties.

Incident 19

Ash Township Volunteer Fire Department, Carleton, Michigan
On February 24th, Firefighter Francis Ploeger arrived at the scene of a two-alarm barn fire. While pulling a hose from the fire truck, he collapsed due to a heart attack. CPR was initiated at the scene, and he was taken to a hospital where he was pronounced dead later that evening.

Incident 20

Allentown Road Volunteer Fire, Fort Washington, Maryland
On March 2nd, volunteer Firefighter Leonardo Maguidad went into cardiac arrest at the station while on duty.

Incident 21

Tomah Volunteer Fire Department, Tomah, Wisconsin
On March 7th, volunteer Firefighter Dennis McGary suffered a fatal heart attack after returning from a house fire. After returning, Firefighter McGary was putting away equipment and preparing firehouse items when the heart attack occurred.

Incident 22

Alexandria Volunteer Fire Department, Alexandria, Nebraska
On March 8th, Firefighter Vinton Durflinger, collapsed and died due to a heart attack after checking out a suspected house fire.

Incident 23

Golden City Volunteer Fire Department, Golden City, Missouri
On March 13th, Firefighter Norman Manka was operating the pump at a grass fire, when he collapsed and died due to a heart attack.

Incident 24

Chesapeake Fire Department, Chesapeake, Virginia
On March 18th, firefighters Frank Young and John Hudgins, died while battling a blaze in an Advanced Auto Parts Store. Both firefighters became trapped by fire when the truss roof collapsed on top of them. The firefighters were found in the rear of the structure some time later.

Incident 25

Granville Fire Protection District, Granville, Illinois
On April 7th, Firefighter Robert Duvall suffered a fatal heart attack while fighting a house fire in Hennepin, IL.

Incident 26

Almena V. Fire Department, Almena, Kansas
On April 8th, Firefighter Norman Adams died from an asthma attack after engaging in support duties for 9 1/2 hours at an industrial fire in a plant that makes aluminum products.

Incident 27

Grant County Fire District 5, Moses Lake, Washington
On April 8th, firefighters Boster and Fowler were responding to a reported mobile home fire when they rounded a corner too quickly and the tanker they were in rolled onto its side. Firefighter Boster was killed and Fowler was treated for broken ribs and other minor injuries. It was not reported whether or not the firefighters were wearing seatbelts.

Incident 28

Harlan Township Fire & Rescue, Pleasant Plain, Ohio
On April 10th, Firefighter Terry Leasher died of internal injuries due to a motor vehicle accident. He was on his way to the station to perform truck inspections.

Incident 29

Schenectady City Fire Department, Schenectady, New York
On April 19th, Firefighter Donald Collins was hooking up a hose at the scene of a vacant house fire around 2 a.m. when he went into cardiac arrest, collapsed, and was taken to the hospital and pronounced dead.

Incident 30

Wayne Fire Department, Wayne, Oklahoma
On April 19th, Firefighter Mathew Hatcher died of abdominal injuries after being pinned between two fire trucks at a grass fire. He was at the rear of one truck starting the pump when a second truck pinned him.

Incident 31

Omaha Fire Department, Omaha, Nebraska
On April 23rd, Firefighter Goessling was killed when the roof collapsed on him at a 4-alarm fire in a commercial building (Dollar General). A 15-yr old has been arrested on suspected arson.

Incident 32

Division of Forestry & Wildlife, Wailuko, Hawaii
On April 23rd, Firefighter Mark Clark died while participating in a chainsaw (tree felling) training class. He was clearing brush around a tree when he put his saw down, collapsed, and died of a heart attack.

Incident 33

Jackson Fire Department, Jackson, Mississippi
On April 24th, Captain Stanley Adams, Captain Don Moree, District Chief Willie Craft, and District Chief Rick Robbins were shot to death during a meeting among district chiefs. They were killed by a disgruntled firefighter who went on a rampage killing five people including his wife. Two other people were also injured in this incident.

Incident 34

Antioch Volunteer Fire Department, Beebe, Arkansas
On April 26th, Firefighter Robert Pemberton was killed in an apparatus accident while enroute to a reported structure fire. He was ejected from the driver's seat after the truck failed to negotiate a turn and then overturned several times. He was pronounced dead at the scene.

Incident 35

Atlanta Fire Department, Atlanta, Georgia
On May 1st, Firefighter Robert Hamler suffered a stroke at the fire station. He had been out inspecting fire hydrants when he started to feel poorly. He was then taken to the hospital.

Incident 36

Mahwah Township Fire Department, Mahwah, New Jersey
On May 25th, Assistant Chief Kevin Malone died of a heart attack at home after returning from a false alarm. He had complained of not feeling well earlier in the day, and two days earlier he had been involved in another fire where he took in a great deal of smoke.

Incident 37

Camp Hill Fire Department, Camp Hill, Pennsylvania
On May 30th, Firefighter William Frank had a heart attack after returning from a heat exchanger fire at a mall.

Incident 38

Globe Ranger District (USFS), Globe, Arizona
On June 9th, Firefighter Michelle Smith disappeared during a training run and was found dead twenty-six hours later. According to the autopsy report, she died of heat exhaustion and dehydration. There was no sign of struggle or foul play.

Incident 39

Cameron Fire Department, Cameron, New York
On June 19th, Firefighter Rex Hoad died from injuries from a motor vehicle accident that occurred while returning from a service call.

Incident 40

Poplar Springs Fire Department, Moore, South Carolina
On June 23rd, firefighters Steele and Harmon were killed when they lost control of their truck, ran off the road, overturned, and hit a tree. The call they were responding to turned out to be a false alarm.

Incident 41

Dillon County Fire Department, Dillon, South Carolina
On June 24th, Firefighter Ronald Lupo was responding to a field fire at approximately 7:10 p.m. While enroute, his vehicle was struck on the right front side by an oncoming van. He was rushed to the hospital and died later that night from internal injuries.

Incident 42

Elizabeth Volunteer Fire Department, Elizabeth, West Virginia
On June 29th, Firefighter Robert Bibbee was hauling drinking water to families and homes in a rural area to raise funds for a fire department event when he suffered a heart attack.

Incident 43

Cameron Volunteer Fire Department, Cameron, West Virginia
Captain Parsons, a liscensed pyrotechnician, was killed at the annual 4th of July Fireworks display that is sponsored by the Cameron Fire Department. A 6-inch round prematurely detonated on the ground causing a piece of metal/wood to strike Parsons in the head. His brother was also injured.

Incident 44

Relief Hose Company No#2, Raritan Borough, New Jersey
On July 11th, Firefighter Bruce Lindner died from cardiac arrest during a vehicle extrication drill

Incident 45

Morgan County Fire Department, Madison, Georgia
On July 11th, Firefighter George Crane Jr. died in a motor vehicle accident while responding in his personal vehicle to a 911 call.

Incident 46

Holyoke Fire Department, Holyoke, Massachusetts
On July 13th, Firefighter Arthur Petit died due to cardiac arrest while searching for victims at a multiple-family dwelling fire. The firefighter's crew was ordered to search the interior of the third floor (fire floor). While searching and ventilating, Firefighter Petit collapsed on the porch of one of the apartments. He was not revived despite the attempts of fellow firefighters who performed CPR immediately.

Incident 47

Pigeon Township Volunteer Fire Department, Dale, Indiana
On July 21st, Firefighter Donald Raibley was responding to a residential house fire at four in the morning when he had a seizure and drove his car over a dam into a lake, where he drowned.

Incident 48

Sligo Volunteer Fire Department, Sligo, Pennsylvania
On July 27th, Firefighter Kris Sherman died from injuries resulting from an overturned pumper during a response to an incident.

Incident 49

Stokes-Rockingham V.F.D, Pine Hall, North Carolina
On July 28th, Firefighter Guyer was responding in his personal vehicle to a transformer fire when his truck hydroplaned on the wet road and collided with an oncoming truck.

Incident 50

Lagro Township Volunteer Fire Department, Wabash, Indiana
On August 6th, Firefighter Swan responded to a motor vehicle accident involving a car and a motorcycle. The driver of the motorcycle ran into the boat that the car was towing. The driver of the car proceeded to shoot the driver of the motorcycle, two bystanders, and the firefighter who arrived on the scene.

Incident 51

Fruitland Fire Department, Fruitland, Utah
On August 8th, Firefighter Norman Ray suffered cardiac arrest due to overexertion at a grass fire.

Incident 52

Addison Fire Department, Vergennes, Vermont
On August 8th, a fire broke out at the barn of the Firefighter Floyd Birchmore. He called the fire department and started to lead the animals out of the barn. The first engine arrived and Firefighter Birchmore began pulling the hose off the truck. Shortly after, he collapsed and died of a heart attack.

Incident 53

Surfside Volunteer Fire Department, Freeport, Texas
On August 8th, Firefighter Mac McGinnis was the only responder to a electrical pole fire that occurred midday. As he began deploying a hoseline, the pole broke in half, bringing

the charged power lines down and electrocuting him. Pole fires occur during drought conditions that allow encrusted salt to cover insulators. There were 25 of these fires previously that summer.

Incident 54

Metal Township Volunteer Fire & Ambulance Company, Fannettsburg, Pennsylvania
On August 12th, Chief Bricker suffered a fatal heart attack while providing patient care on an ambulance crew. Despite the efforts of his own and a neighboring ambulance crew, he was pronounced dead upon arrival at the hospital.

Incident 55

Union Township Fire Department, Union Township, New Jersey
On August 24th, Deputy Chief Leslie Hendricks died as a result of cardiac arrest that was connected to an earlier fire incident at a Burger King. The chief was supervising a crew when a cloud of gas vapors engulfed him. He began having trouble breathing and was sent to the hospital. After staying two days, he returned home and died ten days later.

Incident 56

Gerton Fire Department, Gerton, North Carolina
On August 21st, Firefighter Leonard Coulter died as a result of cardiac arrest while responding to a motor vehicle accident.

Incident 57

Bureau of Land Management, Reno, Nevada
On August 25th, Firefighter John Gray was repairing a water tender between fires when he died of a heart attack. After he collapsed, a fellow crew member began CPR and called an ambulance.

Incident 58

Harahan Fire Department, Harahan, Louisiana
On August 26th, Lieutenant Lawrence Roche had a heart attack at the scene of a structure fire.

Incident 59

South Portland Fire Department, South Portland, Maine
On August 27th, Captain Robert Wallingford died from a heart attack while directing engine company operations at the scene of a four-alarm fire in a welding supply company.

Incident 60

Bahama Volunteer Fire Department, Bahama, North Carolina
On September 4th, firefighters were responding to a call when a tree fell across the roadway and struck the brush truck. The accident killed Firefighter Rick Dorsey and injured one other.

Incident 61

Powell Volunteer Fire Department, Powell, Wyoming
On September 10th, Assistant Chief Bruce Honstain was attempting to rescue his son from a motor vehicle accident when they were both electrocuted and died.

Incident 62

Rising Sun Fire Department, Rising Sun, Maryland
On September 14th, volunteer Firefighter Sam Strall collapsed and died of a heart attack during a fund-raiser at the firehouse.

Incident 63

Baltic Fire & Rescue Department, Baltic, Ohio
On September 18th, volunteer Firefighter Jeffrey Renner had arrived at his regular job, when he was informed of a fire in the paint shed. He was leaving to drive to the station to get his gear when he suffered a heart attack as he was getting to his car.

Incident 64

Springdale Fire Department, Springdale, Ohio
On September 18th, Firefighter Henry Scott suffered a fatal heart attack while at a live burn training exercise.

Incident 65

Birmingham Fire and Rescue Service, Birmingham, Alabama
On September 20th, Firefighter William Reid died of cardiac arrest at required annual fitness test (running & walking). He suffered an acute MI and died after being taken to the hospital.

Incident 66

Herrin Fire Department, Herrin, Illinois
On September 29th, volunteer Firefighter Kevin Reveal died while fighting a fire at a commercial 2-story, vacant, boarded up, wooden building. He was opening up a boarded window that was opposite from where the fire was located when the wall collapsed, killing him and injuring several others.

Incident 67

Mt. Pleasant Rural Fire, Columbia, Tennessee
On October 12th, Firefighter Clark Derryberry died in a motor vehicle accident while returning home from a barn fire. The barn fire was the last one out of a series of four

Incident 68

Jefferson Parish EBC Fire Dept., Jefferson, Louisiana
On October 13th, Firefighter Keith Boudoin was preparing to enter a structure fire for the third time to look for trapped victims when he suffered a fatal heart attack. He was immediately taken to the hospital where he was pronounced dead.

Incident 69

Cowlesville Volunteer Fire Company, Cowlesville, New York
On October 15th, Assistant Chief Karl Schmidt died from an apparent massive coronary after attending an EMS training event. After the event, the chief went immediately to the hospital after arriving at home. The chief died on the way to the hospital.

Incident 70

West Etowah VFD, Altoona, Alabama
On October 18th, Firefighter Martha Ann Bice was cutting firebreaks at a brush fire when she experienced chest pain and collapsed. She was taken to the hospital where they determined that she had a heart attack. While in the hospital they performed a triple bypass surgery. She went home from the hospital and died a couple of days later from complications.

Incident 71

Westminster Fire Department, Westminster, Maryland
On October 19th, Firefighter Eugene Bauerline suffered a fatal heart attack at the fire department. He had been on-duty all morning at the station cooking for the fire department fund-raiser. At about 11:00 am, he left to direct traffic for a college homecoming. This was part of his duty for the "fire police" which is a division of the fire department. He then came back to the fire department for breakfast and later went into cardiac arrest.

Incident 72

Glassy Mountain Fire Department, Landrum, South Carolina
On October 24th, Captain Jackson Capps died in a motor vehicle accident when his fire truck was struck by a dump truck while responding to a grass fire. The Captain had just finished working the third shift at the electric plant when the call for a grass fire came out.

After picking up a fire truck at the fire department, he pulled into a dump truck's path after driving less than 100 yards. The call turned out to be a false alarm.

Incident 73

Cedar Falls Fire Department, Cedar Falls, Iowa
On October 24th, Firefighter Jack Grosse suffered cardiac arrhythmia and died while asleep in his quarters.

Incident 74

Blauvelt Volunteer Fire Company, Blauvelt, New York
On October 26th, Firefighter Albert DeFlumere died of smoke inhalation at a residential structure fire when he returned inside to rescue his son.

Incident 75

Portage Fire Department, Indiana
On October 26th, Firefighter Frank Gilbert died from complications due to MVA while transporting patient to hospital.

Incident 76

Upper Gwynedd Township Fire Department, West Point, Pennsylvania
On November 9th, Firefighter John Bryant was fatally injured in a motor vehicle accident while responding to an alarm. The firefighter was on his way to the station when his vehicle was hit from behind at high rate of speed.

Incident 77

Sharptown Volunteer Fire Department, Sharptown, Maryland
On November 9th, volunteer Firefighter Steve Trice stopped at a motor vehicle accident and was struck by a passing vehicle while attempting to extricate a victim.

Incident 78

Tonawanda Fire Department, Tonawanda, New York
On November 12th, Captain Walter Schwinger Jr. died of a pulmonary embolism while asleep in the bunkroom while on-duty. The crew received a call and went back in to check on the Captain when he did not get up.

Incident 79

Anne Arundel County Fire Department, Millersville, Maryland
On November 12th, Firefighter William Chambers collapsed and died of a heart attack during response to medical call.

Incident 80

Highview Fire District, Louisville, Kentucky
On November 24th, Firefighter Donald Manuel suffered a fatal heart attack upon arrival at the scene of a church fire.

Incident 81

Branford Fire Department, Branford, Connecticut
On November 27th, Firefighter Edward Ramos was killed in a warehouse fire at Floors and More, Inc. after the roof collapsed, trapping him and two other firefighters inside. Despite having his SCBA facepiece knocked off in the collapse, Firefighter Ramos stayed on the line and knocked down the fire so his comrades could escape.

Incident 82

Houston Volunteer Fire Department, Houston, Texas
On December 4th, District Chief Ruben Lopez was killed in a residential structure fire while attempting to rescue one of the house's occupants. The firefighter and the victim were caught in a flashover. Both the firefighter and the victim were killed.

Incident 83

Somers Fire Department, Somers, Connecticut
On December 8th, Firefighter Craig Arnone was electrocuted when his SCBA tank came into contact with a downed power line carrying 23,000 volts at a residential structure fire. A snowstorm was responsible for the power line being down. Firefighters thought the electrical power to the area was shut-off when they came to the house.

Incident 84

Chicago Fire Department, Chicago, Illinois
On December 21st, Firefighter Stanley Scott suffered a fatal heart attack after hooking up a hydrant at the scene of a structure fire. CPR was initiated on the scene, but firefighter Scott was pronounced dead later at a hospital.

Incident 85

Boston Fire Department, Boston, Massachusetts
On December 21st, Firefighter James A. Ellis died as a result of injuries sustained after falling approximately 20 feet down a fire pole on the way to a call. The presence of water from possibly a sink is listed as the cause of the fall. The fall caused severe head trauma and neurological damage.

Incident 86

Stroh Volunteer Fire Department, Stroh, Indiana
On December 21st, Firefighter Laura Halsey was driving to the hospital with a patient from an automobile wreck, when a car struck them head on around 4:30 am (no headlights and in wrong lane). All the occupants inside the car were killed.

Incident 87

Burnet Volunteer Fire Department, Burnet, Texas
On December 23rd, Firefighter James Warick was struck by a vehicle while directing traffic at an incident.

Incident 88

Forsyth County Fire Department, Cumming, Georgia
On December 27th, Firefighter Chesney was killed while advancing a hoseline to the upper floor of a three-story condo fire when the roof collapsed due to unseen fire spread. The two firefighters with him were able to escape.

Incident 89

Kimball Township Fire Department, Hurley, Wisconsin
On December 29th, Chief Raymond Emmrich suffered a heart attack while responding to a dwelling fire. At the time of the heart attack, he was driving a pumper. He went about a block before driving into a snow bank